Under New England

Under New England

The Story of New England's Rocks and Fossils

Written and Illustrated by Charles Ferguson Barker

University Press of New England
Hanover and London

Published by University Press of New England,
One Court Street, Lebanon, NH 03766
www.upne.com

Printed in Singapore
5 4 3 2 1

Under New England: The Story of New England's Rocks and Fossils is an adaptation of *Under Michigan: The Story of Michigan's Rocks and Fossils* by Charles Ferguson Barker and is published by arrangement with Wayne State University Press.

Library of Congress Cataloging-in-Publication Data

Barker, Charles Ferguson.
Under New England : the story of New England's rocks and fossils / Charles Ferguson Barker. — 1st ed.
p. cm.
ISBN-13: 978-1-58465-696-8 (cloth : alk. paper)
ISBN-10: 1-58465-696-4 (cloth : alk. paper)
1. Geology—New England—Juvenile literature. 2. Geology, Stratigraphic—Juvenile literature. 3. Fossils—New England—Juvenile literature. I. Title.
QE78.3.B27 2008
557.4—dc22 2007044689

For Clare

Acknowledgments

First, I would like to thank University Press of New England for accepting this project—specifically Dr. Phyllis Deutsch, Richard Pult, and others at the Press who first reeled in the idea from my initial inquiry. Richard convinced me that the work might cover all of New England rather than the individual states. He advised, "Don't let the political boundaries of the region force your hand." Good advice on many levels, especially geology.

Thanks to my mother and father for giving me books when I was growing up that I loved to read (and still do), including the works of Maurice Sendak and New England's own Robert McKloskey. My grandmother, Ms. Kate Ferguson, gave me encouragement to draw, paint, and write, and keep looking, and looking.

On the New England geology side of things, I'd like to thank the following for their review and comments on specific sections of the draft manuscript:

Dr. Duncan M. FitzGerald, Boston University: coastal and marine geology

Dr. Brendan Murphy, St. Francis Xavier University: tectonic history of the region

Dr. Thomas Weddle, Maine Geological Survey: glacial geology and hydrogeology

Their expert input was extremely valuable. Still, any errors in this book are mine alone.

Finally, thanks to my wife, Peg, and our wonderful daughter, Clare, for their support, understanding, and encouragement.

Author's Note

The geology of New England is, in a word, complex. The region's geologic history is hard to decipher and describe, and new ideas about what happened, and when, are coming up all the time. As D. W. Caldwell put it so well in his book *Roadside Geology of Maine*, "New ideas in geology are often controversial, especially when they appear in print." Forming a geologic history is like putting together a jigsaw puzzle. In the case of New England, it is as if the pieces, before being put together, were left out in the rain, became curled and faded, and then the dog chewed on them. That said, I have tried to piece together a basic story by reviewing many sources of literature on the subject, studying geologic maps, and partaking in the usual amount of arm-waving with fellow geologists, which both clarifies the subject and sometimes muddies it, but is perhaps the most enjoyable activity of our field. I hope I have got it correct enough to satisfy most geologists.

My goal in this book and others like it is to take readers who may already be familiar with a place in its horizontal dimension and show them some not-so-familiar views in the vertical direction—or what's "under" that place. And to tell the story of its formation throughout the distant past. To show and tell, that is, a geologic history—one that takes readers back in time to see the fascinating, often surprising transformations that, however unlikely such events seem now in this familiar place, yet did occur, and that ultimately created the landscape they now know. Maybe, by looking at the place in this way, we all might be surprised and perhaps appreciate the place even more.

My further hope is that the quotations from various New England poets provide a link from the geology of the land to the literature of the land. For if geology is the anatomy of a region, and poems about that region are an impression of it, why not put them together to more completely define a sense of place? One particular phrase from a poem I found struck a chord in me after returning from a trip to New England while working on this book—I had been scrambling over rocks in Ogunquit, Maine, the same rocks I had scrambled over twenty-five years ago when I was much younger. I noticed the rocks were the same, and there was comfort in that. Perhaps some adults reading this book might relate to the quote, and maybe, much later on, children who read this book might relate to it as they grow older:

> *. . . I shall find the sullen rocks and skies*
> *Unchanged from what they were when I was young.*
> *—Edna St. Vincent Millay*

It seems Henry David Thoreau had similar feelings. In *Walden* he writes of his beloved Walden Pond:

> it is itself unchanged, the same water which my youthful eyes fell upon; all the change is in me.

And finally, I believe that learning about the geology of a place is good for you. Perhaps this idea was best expressed in 1832 by a reviewer of Scottish geologist Charles Lyell's book *Principles of Geology* (Lyell spent some time in New England studying its geology). The reviewer wrote:

> There are other investigations which more nearly affect our social happiness than the philosophy of geology, but perhaps there is none which in an indirect manner produce a more

> wholesome and beneficial effect upon the mind. . . . After the perusal of Mr. Lyell's volume, we confess to emotions of humility, to aspirations of the mind, to an elevation of thought, altogether foreign from the ordinary temper of worldly and busy men. . . . So disposed mentally, the heart overflows with charity and compassion; vanity shrivels into nothingness; wrongs are forgotten, errors forgiven, prejudices fade away; the present is taken at its real value; virtue is tried by an eternal standard. There are sermons in the stones and tongues in brooks, but they want an interpreter: that interpreter is the enlightened geologist. Such a man is Mr. Lyell.

I do not claim to be an enlightened geologist, but I have known a few. Dee Caldwell of the Department of Geology at Boston University was one.

The following references were useful in writing this book, and are some of the many good sources of information on New England Geology:

- *Roadside Geology of Maine*, by D. W. Caldwell, Mountain Press, 1998
- *Written in Stone: A Geological History of the Northeastern United States*, by Chet Raymo and Maureen Raymo, Black Dome Press, 2001
- *The Teacher-Friendly Guide to the Geology of the Northeastern U.S.*, by Jane Ansley, Paleontological Research Institution, Ithaca, N.Y., 2000

Numerous technical papers and conference proceedings were also of great use in preparing the manuscript. Perhaps of greatest reference and inspiration were the courses and interactions with the people of Boston University's Department of Geology, especially Duncan M. FitzGerald, Dee Caldwell, and Jonathan Lincoln.

Under New England

This is the place. Stand still, my steed,—
Let me review the scene,
And summon from the shadowy past
The forms that once have been.
—Henry Wadsworth Longfellow
(New England poet, 1807–1882)

15,000 years ago

160,000,000 years ago

450,000,000 years ago

You know New England.
Know its old and familiar landscape.
Know Vermont's Green Mountains and colorful woods,
New Hampshire's blue skies and White Mountains,
Know Massachusetts from the Berkshires rolling down to Boston.
You know New England's Cape Cod and the crowded sandy beaches of its islands,
Or the quiet Maine woods and its wild remote rocky coast,
Know Rhode Island's quiet bays and islands,
And Connecticut with its winding rivers and old stone walls.
You know New England's warm muddy springs,
its hot sticky summers,
the crisp cool bright-leafed days of autumn,
And the cold snowy winters when the sun sets much too
soon in the afternoon.

You may know all about what covers New England,
But have you ever wondered what's under New England?

Under New England there are all kinds of rocks and fossils that tell a fascinating story:

There are rocks from the floors of ancient seas before there was an Atlantic Ocean.

There are rocks from fiery volcanoes that erupted fountains of lava across the surface of the land, and rocks from molten magma that cooled slowly deep underground.

There are rocks from mile-thick icy cold glaciers that dropped sand, gravel, boulders, and clay to create Cape Cod, Nantucket, and Martha's Vineyard.

There are rocks from when continents crushed together and folded over older layers of rock, like when a carpet slides on a polished wooden floor and crumples up against a wall. And there are rocks and fossils from when dinosaurs roamed across old New England.

There are rocks with funny names like batholiths and laccoliths, and there are xenoliths and lopoliths. Some people may think the rocks of New England are all pretty much the same, but don't tell that to a geologist!

This book tells a history of New England, but not of Pilgrims or Patriots or colonies. This is a story of New England's rocks and fossils—a history not of hundreds of years ago, but of hundreds of millions of years in the past.

This is a story of colliding continents and fiery volcanoes on chains of islands that moved far across ancient oceans to build the land we now know as New England.

Heaven is under our feet, as well as over our heads.
—Henry David Thoreau (New England naturalist/writer, 1817–1862)

You could say that the history of the rocks and fossils of New England is like the first motto of the United States: "E Pluribus Unum." It is written on every U.S. coin or dollar bill—go ahead and look for yourself!

E Pluribus Unum are words from the Latin language that mean "From Many, One." The nation's motto reflects the uniting of many independent colonies and also the many different types of people who have come together to make one nation.

The motto "From Many, One" also applies to the story of the rocks and fossils of New England because the land under New England was formed by many different small continents and volcanic island chains that collided together over hundreds of millions of years to create one land—New England (and much of the rest of the East Coast of the North American continent too!)

From many

. . . one

But where did these ancient rocks and fossils come from?
And how can continents and volcanic island chains move across oceans?

Let's take a look.

The hard outer shell of the earth is broken up into huge plates of rock, like the cracked shell of a hard-boiled egg. The earth's rigid plates slowly glide over a layer of softer rock below, a bit like a raft that drifts on a lake.

The plates of rock covering the earth are slowly moving, floating over a layer of weaker rock, and have been drifting for billions of years. Even as you read this book, the continents and ocean floors are drifting along. But they move very gradually—about as slowly as your fingernails grow!

Even though the plates move slowly, over hundreds of millions of years entire continents can travel around the world, and oceans can open and close like a clamshell.

New England was formed from those huge plates of the earth sliding around like blocks of ice in a frozen stream. The continents change position and shape on the globe like big puffy clouds that slowly change when you watch them in the sky.

Because the continents move around the earth, they slowly pass through areas of different climates. New England was located near the equator nearly 400 million years ago, and we can find fossils in the rocks of New England that lived in a climate that was warm and tropical. Those rocks and fossils tell us that many parts of New England were once south of the equator!

The huge plates of rock slowly grind into each other, or pull apart from each other, or slide past one another—all these things happened to the rocks under New England a long time ago. Sometimes, we feel earthquakes in New England as the old plates shift a little. But there are not nearly as many earthquakes in New England as there are in California, where big plates are still colliding and slipping against each other!

Sometimes the plates of rock of the outer shell of the earth split apart and oceans open up. Molten rock pushes up at the center of huge cracks (called rifts) as the land pulls apart. The Atlantic Ocean formed in this way about 200 million years ago, and the Red Sea is beginning to open up in the same way today.

Many plates are mostly made up of oceanic crust of the seafloor, which is heavier than the rocks of continents. Some plates are made up of both oceanic crust and continental crust. In that case, the continents are embedded in the plate like peanuts in peanut brittle (they are BIG peanuts), and are surrounded by oceanic crust, and the entire plate moves as one.

Because some plates of the earth have continents riding along with them, you could say the plate is like the conveyor belt you see at a grocery store checkout lane, and the continents are like groceries riding with them toward the cash register. Have you noticed what happens at the end of the conveyor belt? At the grocery store, the belt dives down under the edge of a solid platform. On the earth's surface, a plate of oceanic crust does the same thing when it meets up against another plate.

If the grocery store's conveyor belt keeps moving, the groceries on it will all collide at the end of the lane where the belt dives down (especially if the checkout person isn't paying attention). In sort of the same way, the continents riding on a plate made mostly of oceanic crust will pile up against another plate while the oceanic crust dives under it, like a store's conveyor belt. That's because rock of the continental crust is generally lighter than the ocean floor and does not get pulled down with the diving seafloor.

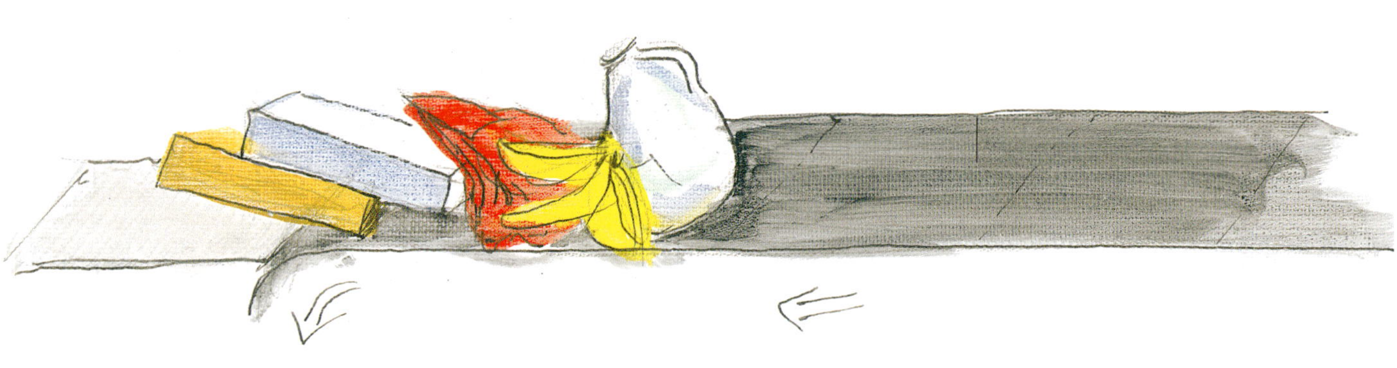

When a moving oceanic plate collides with a continent (or with another oceanic plate), the heavier oceanic plate dives deep down into the earth, where it is hot, and melts. That melted rock, called magma, rises up toward the surface because it is very hot. (You may have learned that warm air rises; so does warm molten rock). If the magma cools and crystallizes deep under the surface, it can form granite or other hard rocks. If the molten rock makes it up onto the surface, it erupts in fiery volcanoes and cools into volcanic rock.

Most of the land of New England was added to the ancient coast of the North American continent over hundreds of millions of years, beginning about 500 million years ago. The plates pushed smaller slabs of land (called micro-continents), chains of volcanic islands, and the muddy floors of ancient seas into the growing East Coast like a sculptor adds slabs of clay onto a sculpture.

If we think of New England as forming like a jumble of groceries piling up at the end of a checkout lane (it's actually much more complicated than this), the western part of New England, where the Berkshires and other mountains rise up, is like a layer cake that crumpled at the end of the conveyor belt. This happened about 500 million years ago when an ancient ocean floor was colliding with the East Coast of North America.

Over time, more and more land was added to New England like groceries moving down a checkout lane. All this took hundreds of millions of years—this was not the express lane!

The central part of New England is like a bag of Maine potatoes on a grocery store conveyor belt that rammed into the layer cake. This happened about 450 million years ago as a landmass called Ganderia, made up of muddy seafloor and volcanic islands, crashed into the East Coast.

Most of eastern New England is like a big can of Boston baked beans on a conveyor belt that collided with all the other groceries in front of it as the conveyor belt kept moving. This happened about 400 million years ago as a landmass called Avalonia smashed into the East Coast.

About 450 million years ago, if you were standing
at the eastern coast of North America,
you would see mountains nearby to the west,
and volcanic islands approaching from the east.

All I could see from where I stood
Was three long mountains and a wood;
I turned and looked the other way
And saw three islands in a bay.
—Edna St. Vincent Millay
(New England poet, 1892–1950)

While Avalonia was being pushed into North America, there was a giant CRUNCH as all the groceries (the ancient seafloor, volcanic islands, and micro-continents of New England) got squeezed together when yet another package moved down the conveyor belt and rammed into them. That huge package was the northern part of the early continent of Europe. First it collided to the north around Greenland, and then with New England, squeezing everything in between and pushing up the northern Appalachian Mountains. Later, south of New England, the continent of Africa collided with the rest of the East Coast to form the southern Appalachian Mountains. This happened about 300 million years ago and made one huge super-continent called Pangea, with New England at its center!

After all that crushing together of continents, something else happened. The conveyor that had pushed continents together in one direction now reversed and started moving the other way—so the continents of Europe and Africa started pulling away from the East Coast of North America. The basin that opened up between them became the young Atlantic Ocean as the plates kept spreading apart!

Just before the Atlantic Ocean began to open, the land cracked and started to split apart in many places. The Atlantic didn't open all at once or all in the same place. As the land pulled apart, some places cracked and sank down like the stretching coils when you pull apart a Slinky® toy. Massachusetts split, and Connecticut cracked. That's why the Connecticut River Valley is where it is!

Finally, the young Atlantic Ocean found a place east of New England where it could continue to split open, and the huge continent of Pangea began to break apart. As the Atlantic opened, it separated northern Europe from New England, and Africa separated from the rest of North America to the south—and the ocean is still widening!

When the Atlantic Ocean opened, it split apart the older rocks that had once been together. This is why the rocks under South Boston today are the same as the ones under southern Ireland. Three hundred million years ago, it wasn't such a long way to travel from New England to Ireland. In fact, you could have walked.

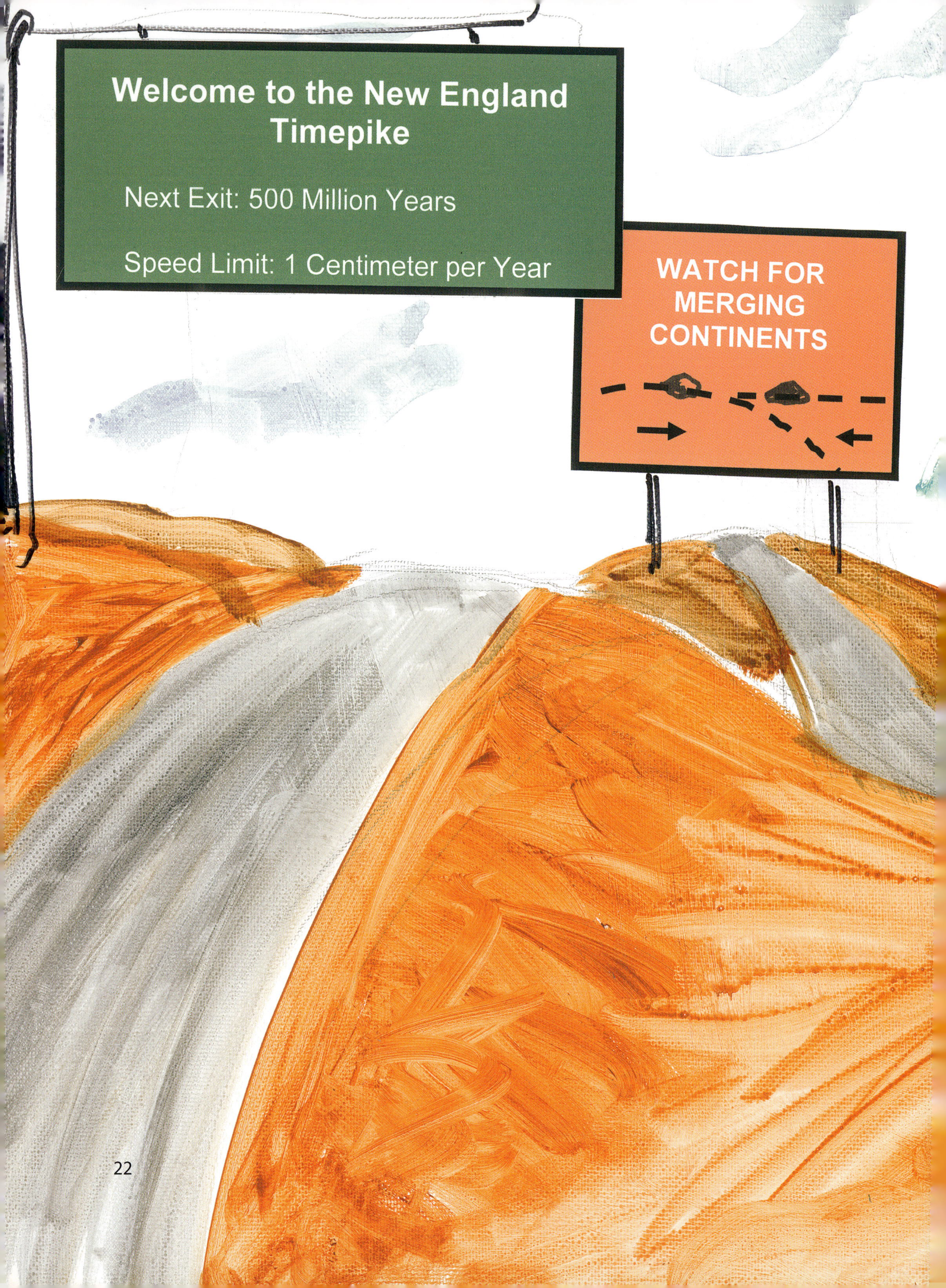
Welcome to the New England Timepike
Next Exit: 500 Million Years
Speed Limit: 1 Centimeter per Year
WATCH FOR MERGING CONTINENTS

Although New England was formed like a bunch of groceries colliding at the end of a checkout lane, New England is not made up of cake, potatoes, or baked beans. It is made of rock. And where plants and animals lived on those rocks, they sometimes left fossils.

So, after all that geologic drama, what kinds of rocks and fossils are under New England? There are all three main types of rock: igneous, sedimentary, and metamorphic.

Igneous rocks are formed by magma (molten rock) that cools and solidifies into a hard rock. Igneous rocks can form as magma cools slowly deep underground, or if magma erupts onto the surface and cools quickly.

The igneous rocks that cool slowly underground (sometimes taking a hundred thousand years or more to cool!) grow large mineral crystals because they have so much time to crystallize. We can see those large crystals in igneous rocks like granite.

We can see lots of granite in New England, but why is granite at the surface if it cooled slowly deep underground? It's there because wind and rain and ice broke up and washed off the rock above it as the land was slowly rising. Now we see huge mountains of granite at the surface in places like Mount Desert Island in Maine, the White Mountains of New Hampshire, and many others.

The igneous rocks that erupt onto or near the surface cool quickly and so they have small mineral crystals because there was not enough time for the crystals to grow very large. Most volcanic rocks in New England look like big blocks of hard rock with tiny crystals you can barely see.

We can see lots of volcanic rock in New England because much of the land was formed by ancient volcanoes, and we can see rocks formed when magma squirted up through cracks in the crust when the Atlantic Ocean was opening.

In New Hampshire, there is a mountain made of granite called Cannon Mountain. A long time ago some rocks near the top crumbled and fell down, and after that the high cliff looked like the side of an old man's face. It was so interesting that people called it the "Old Man of the Mountain" and put it on the state's road signs and even license plates. But in May of 2003 more rocks crumbled and fell down the mountain. These were the rocks that made up the Old Man. Even though they have tumbled down, someday more rocks will fall off that same cliff, and it will look like something else again.

Just like the shapes you can see in clouds, you can see amazing things in the cliffs and shadows of a mountain—if you look hard enough and use your imagination! Hour after hour, as the angle of the sun shining on the mountain keeps changing, it will make different shapes of the shadows.

. . . Yonder ragged cliff
Has [a] thousand faces in a thousand hours.
—Ralph Waldo Emerson (New England poet, 1803–1882)

Because parts of New England were once the floors of ancient seas, there are lots of sedimentary rocks under New England too.

Sedimentary rocks form by other rocks breaking up into small pieces called sediment that later get cemented together. They can also form when seawater evaporates, leaving behind dissolved minerals called "evaporites." Rock salt and gypsum are evaporite sedimentary rocks. Organic sedimentary rocks are formed by old plants being buried and squeezed—coal is an organic sedimentary rock.

Most of the sedimentary rocks we find in New England formed in ancient oceans or shallow seas. Tiny particles of sediment, washed from the land by rains and rivers, settled in these shallow seas and piled up at the bottom. More and more sediment piled on top and stacked up in thick layers. The oldest layers were on the bottom, just like clothes that pile up on the floor of a messy room. Over time, the sediments squeezed together under all that weight. Eventually some of the liquid surrounding the tiny sediment grains dried up, the particles became glued to each other, and the whole bunch of once-loose sediment turned into solid rock (sometimes as hard as a sidewalk).

By looking at these rocks, we can tell a lot about what the environment was like a long time ago when the sediment was deposited. Usually, if sediment was deposited in calm water, we find a rock with very fine grains because the water wasn't moving fast enough to carry larger sediment grains. Rocks that form from small particles laid down in calm water include mudstone and shale. Marl is a rock made from muddy sediments. If the water was moving faster, we find sedimentary rocks with larger grains, like sandstone. If the water was *really* moving, we can find rocks with gravel and cobbles cemented together, called a conglomerate. (Because of how it looked, early New Englanders called this rock "pudding stone.")

Limestone is a sedimentary rock that often forms in shallow seas. Because of that, we can sometimes find fossils (evidence of former life) in limestone—shells of extinct creatures, coral, and fish teeth and bones. Limestone is sometimes used as building stone, so if you see a gray or tan rock used as blocks for a building, go up close and look at it. If it's limestone, you just might find some fossil shells or coral from an ancient sea!

We can even find fossilized dinosaur footprints in some New England rocks. When the Atlantic Ocean tried to open up around central New England, a big valley formed and then filled with sediment. This valley is where the Connecticut River now flows. Around 200 million years ago dinosaurs walked across the wet muddy land and left footprints. We can still see some of those footprints in the sediment that hardened into rock. Maybe, millions of years from now, your footprints, or your dog's, might become fossils!

Metamorphic rocks form when existing rocks change by intense heat and pressure (*metamorphose* means "to change," as when a caterpillar metamorphoses into a butterfly). Any existing rocks can change into metamorphic rocks by being squeezed, or being next to hot molten magma. Because New England formed from colliding continents and volcanic islands, and melting seafloor created a lot of rising molten rock, you can bet there are lots of metamorphic rocks under New England!

If hot magma pushes up against an existing rock, the heat from the magma can scorch or bake the cooler rock where it makes contact. This is called "contact metamorphism" and usually affects only the nearby rocks.

When whole continents collide, the heat and pressure can metamorphose very large areas of rock. This is called "regional metamorphism." The amount of metamorphism is usually greatest where there is the most heat and pressure. For instance, if two continents collide, and one has a part that sticks out a bit toward the other, that place will likely have the most pressure put on it, and the rocks there will be metamorphosed more than in other places farther away. This is sort of like what happens if you accidentally bump your leg on the sharp edge of a desk or tabletop—where the edge hit your leg is where the bruise will be biggest, and where it will hurt the most!

Time and thought were my surveyors,
They laid their courses well,
They boiled the sea, and baked the layers
Of granite, marl, and shell.
—Ralph Waldo Emerson

We can see lots of metamorphic rocks all over New England. That's because, with all those colliding continents, New England was really crunched!

Maine was strained,
New Hampshire was nudged,
Vermont was squeezed in a vice,
Massachusetts was mashed,
Connecticut was crushed,
And Rhode Island was rammed.

In many places in New England you can see rocks that look all twisted, stretched, crunched, and heated so much they almost melted! Mount Monadnock in New Hampshire is made up of twisted and folded metamorphic rocks, and so are the Green Mountains of Vermont.

Marble is a metamorphic rock created by heating and squeezing limestone. We can find lots of marble in many places in New England. There are marble quarries in western Massachusetts and in Vermont where the rock is cut from the ground and used for buildings and floors.

Long after New England formed from colliding volcanic islands and continents, the surface of the land wore down like a sand castle in the rain. Huge blobs of granite, with large crystals of minerals that had cooled slowly deep underground, are now seen at the surface. Mount Katahdin in Maine is one of these mountains of granite, and there are many others all across, and under, New England.

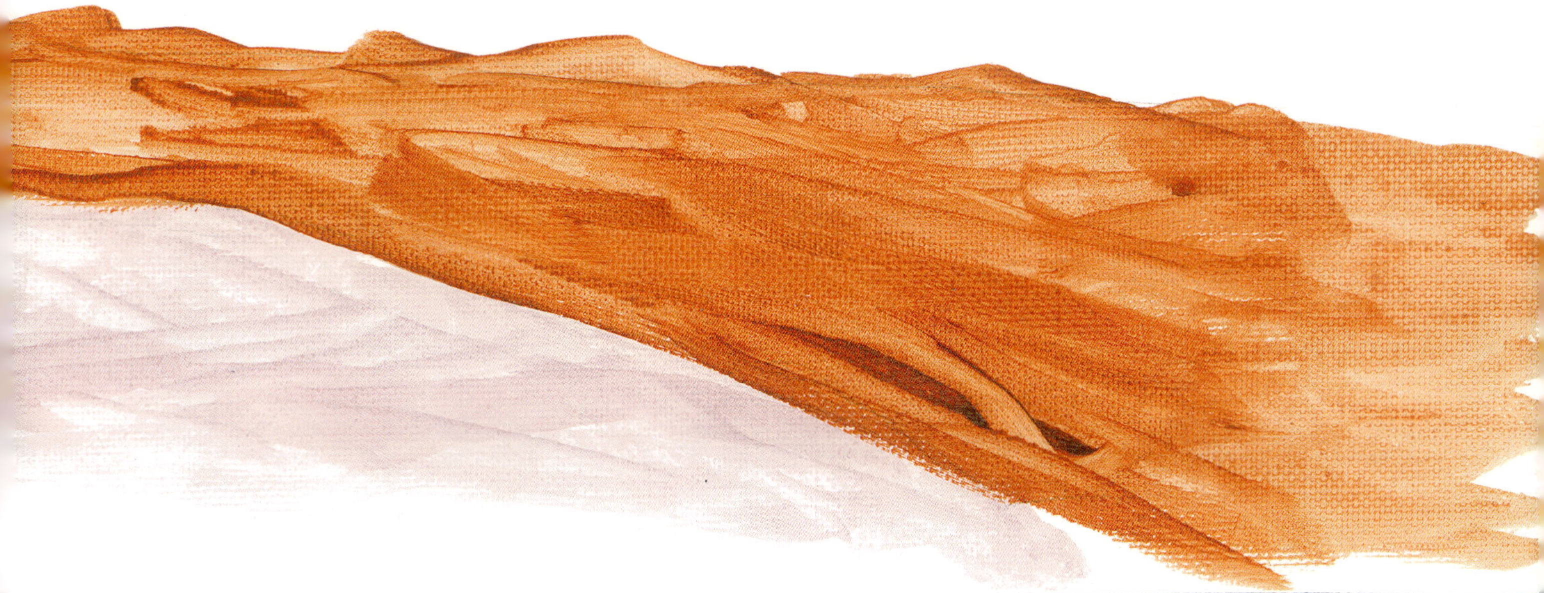

Then, about 2 million years ago, as the climate became colder and colder,

it snowed,
and snowed,
and it snowed.

It was so cold that all the snow that fell during the long, cold winters didn't melt during the short, cool summers. Up in Canada, the snow piled up higher and higher and squeezed into massive slabs of ice called glaciers. The glacial ice, a mile or more thick, grew even larger, moved down from Canada, and covered all of New England. As the glaciers advanced to the south, the climate changed from colder to warmer, and the glaciers melted back. Then the climate changed again, from warmer to colder, and the glaciers moved south again. Back and forth, back and forth—this happened many times!

Like the actions of both a bulldozer and a conveyor belt (moving ice and rock to the front of the glacier), the big thick slabs of ice scrapped across the land, plucking up rocks, grinding rocks, and scratching rocks. The most recent glaciers began pushing down into southern New England and beyond about 24,000 years ago and started melting back about 20,000 years ago. When they melted, they dropped the rocks they carried, and the rocks they had crushed, and left piles of boulders, sand, gravel, silt, and clay.

Long ago,
In the deer-haunted forests of Maine,
When upon Mountain and plain
Lay the snow, . . .
—Henry Wadsworth Longfellow

Some of the places created by these glaciers are Cape Cod, Nantucket, Martha's Vineyard, and Block Island. They are all huge mounds of rock and sand and clay that the glaciers dumped at the end of their bulldozed paths. Where the glaciers pushed down even farther south, they left a long ridge of sand, gravel, and clay called a moraine that forms Long Island in New York. In addition, after melting back toward Canada, they left behind a new sculpture of freshly carved hills and valleys all over the New England landscape.

But before the glaciers melted away, sea level was much lower than it is today. That's because some of the water that's now in the oceans was locked up as ice on the land. It had not yet melted to flow into the sea.

So a long time ago, maybe 20,000 years or so, Cape Cod, Nantucket, and Martha's Vineyard were all part of the mainland—high and dry. You would have had to walk about a hundred miles from any of those places to get to the beach! And when you got there, you would probably have seen icebergs floating by.

As the glacial ice melted, the water it released escaped from the land and flowed back to the sea. The sea level rose, and water surrounded the islands and Cape.

Dear old Nantucket's Isle of Sand,
An ancient exile from the land, . . .
—Frank Dempster Sherman
(American poet and writer, 1860–1916)

The glaciers did something else to New England that may seem surprising—they were so heavy they pushed down the crust of the earth!

Just as a lobster boat pushes down lower in the water when it is filled with big and heavy lobsters, the thick and heavy glaciers, when they covered the land for many thousands of years, pushed down the crust of the earth under New England.

And when the glaciers melted away, and all that weight of their ice was gone, the land started rising back up! Just as, when the big, heavy lobsters are removed from the boat, it rises back up in the water.

The land in Maine has risen more than 400 feet over the last 12,000 years!

When the climate warmed and the glaciers melted, all that water eventually flowed into the ocean. Before the land that the heavy glaciers had pushed down could spring back up, the rising sea moved in over the land, covering much of the present coast of New England. Up in Maine, where the land was pushed down the most, the rising sea could move up old river valleys easily, and the seawater crept many miles inland from where the coast is now.

The sea left thick layers of mud and gray-blue clay with fossil clams and other shells—soft sediments that didn't have enough time to harden into rock. This clay marks the spots where the seawater reached far inland before the land could rise back up again. We can find this thick clay in many places along the New England coast.

It snowed in spring on earth so dry and warm
The flakes could find no landing place to form.
Hordes spent themselves to make it wet and cold,
And still they failed of any lasting hold.
They made no white impression on the black.
They disappeared as if earth sent them back."
—Robert Frost (New England poet, 1875–1963)

Under Boston, there is a layer of blue-gray clay about a hundred feet thick deposited at the time when the sea moved in over the land. The big buildings need to have deep foundations that reach through the slippery mud into solid rock, because clay is too slippery and squishy to support a tall building!

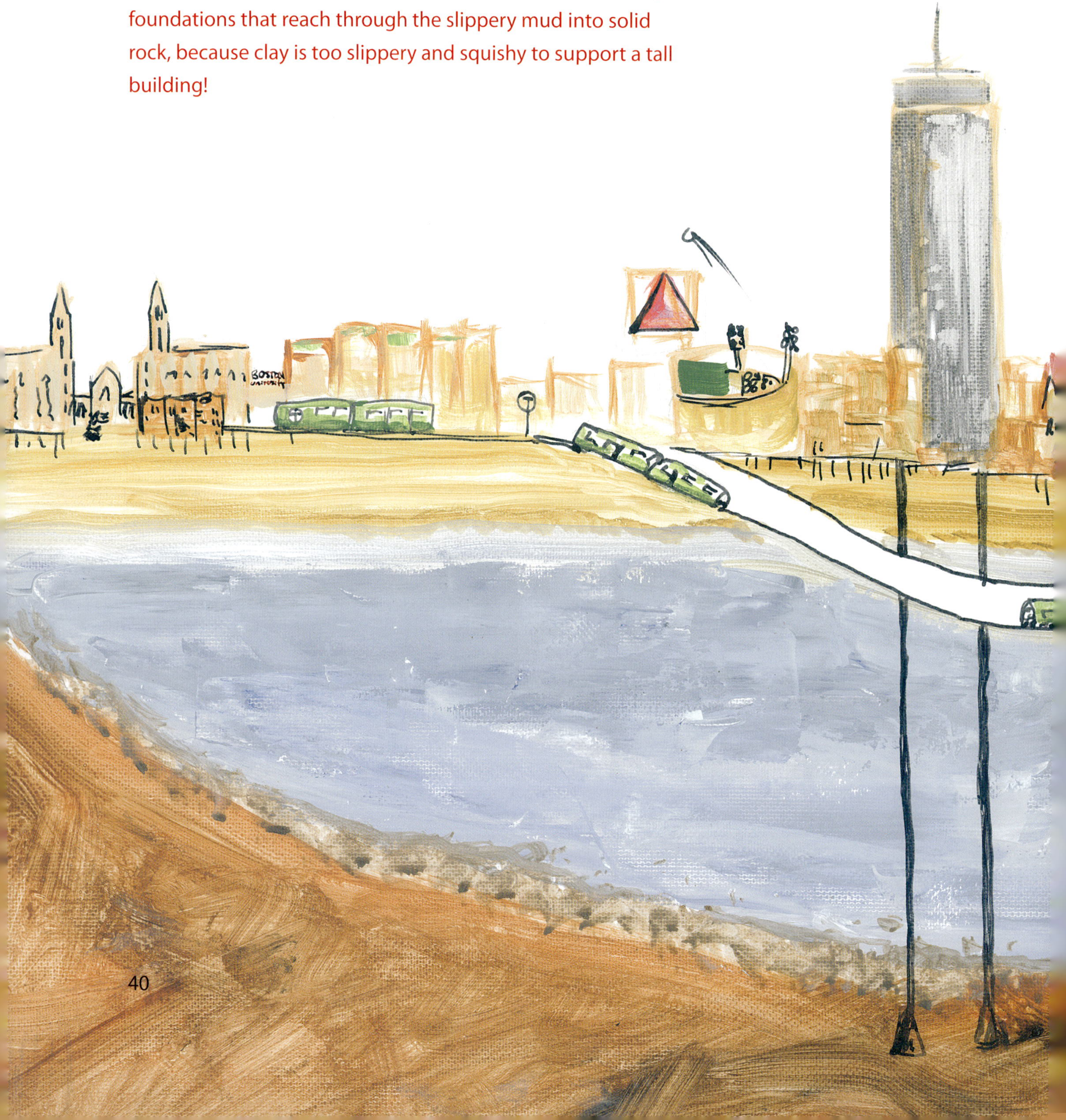

The underground tunnels of the T (Boston's subway), and the "Big Dig," are dug into that same thick clay.

Most of New England's soil is made from ground-up bits of rock left by the glaciers. In some places, the glaciers completely scraped away any earlier soil, leaving only bare rock.

Soil is made of broken-up bits of rock and also bits of old plants called organic matter. This organic matter makes good, rich soil for planting—good for growing potatoes, and corn, and of course lilacs.

Lilacs,
False blue,
White, Purple,
Color of lilac,
Heart-leaves of lilac all over New England,
Roots of lilac under the soil of New England,
Lilac in me because I am New England,
Because my roots are in it,
Because my leaves are of it,
Because my flowers are for it,
Because it is my country
And I speak to it of itself
And sing of it with my own voice
Since certainly it is mine.
—Amy Lowell
(New England poet, 1874–1925)

When rainwater seeps into the soil and soaks deeper into the ground, it fills up the tiny spaces between the sand and rock and becomes groundwater. Many people in New England get their drinking water from wells that go down to the top of the water-soaked ground, which is called the water table. There are no underground streams or lakes under New England; the water is just soaked into the sediment like the water in a sponge. If the hard, solid rock called bedrock has cracks in it, the groundwater can flow into the cracks. Some wells get their water from inside the fractured bedrock.

It's important to remember that almost anything spilled on the ground can seep down into the groundwater and make its way into a drinking-water well. So remember, if it's something you wouldn't drink, don't pour it onto the ground!

The glaciers left many kinds of interesting landforms all over New England. Some of New England's most familiar places, like Beacon Hill in Boston, are features left by the glaciers.

Drumlins (from the Irish word for "hill") are long hills that are steeper on one end than the other (the steep side faces the direction that the glacier came from). Beacon Hill is a drumlin, as are most of the Boston Harbor Islands. All across New England we can find drumlins.

Eskers are the remains of drainage channels that formed underneath glaciers as ice melted. When the glaciers finally melted away, the sand and gravel that was in these channels remained above the ground. Eskers are now seen as long, snake-like ribbons of sand and gravel. They are like river channels, but instead of being cut *into* the ground, eskers are long mounds *on top of* the ground. Because they were high and dry ridges that were sometimes surrounded by swampy land, they made good places for paths and roads for early New Englanders.

A kettle pond is a lake formed when a large chunk of glacial ice is buried under sand and gravel washing down from a melting glacier nearby. When the buried ice finally melts, it leaves a big hole in the ground that can fill with water. Walden Pond in Concord, Massachusetts, is a kettle pond and is where the naturalist Henry David Thoreau spent two years in the woods in a small cabin he built to get away from the hustle and bustle of city life. He wrote down his thoughts in a book titled *Walden*.

> *Walden is blue at one time and green at another, even from the same point of view. Lying between the earth and the heavens, it partakes the color of both.*
>
> *Why— here at Walden; . . . it is the same liquid joy and happiness to itself and its Maker, aye, and it is to me.*
>
> *—Henry David Thoreau*

Although most of New England took hundreds of millions of years to form, you can see places on the New England coast that change after one nor'easter storm.

(A "nor'easter" is a big storm and is named for the strong winds that blow into New England from the northeast. Smaller, less-powerful storms are called "gales.")

The shattered water made a misty din.
Great waves looked over others coming in,
And thought of doing something to the shore
That water never did to land before.
—Robert Frost

The tides rush into the harbors and channels of the New England shore twice a day, and then flow away again. The rising, incoming tide is called the "flood tide"; the falling, outgoing tide is called the "ebb tide." The tides also rise and fall along beaches, but do not produce a current. You might find that a blanket left on the beach during low tide will get covered with water as the tide moves in! Some boats at anchor will touch bottom during low tide and will need to wait until high tide before they can sail away.

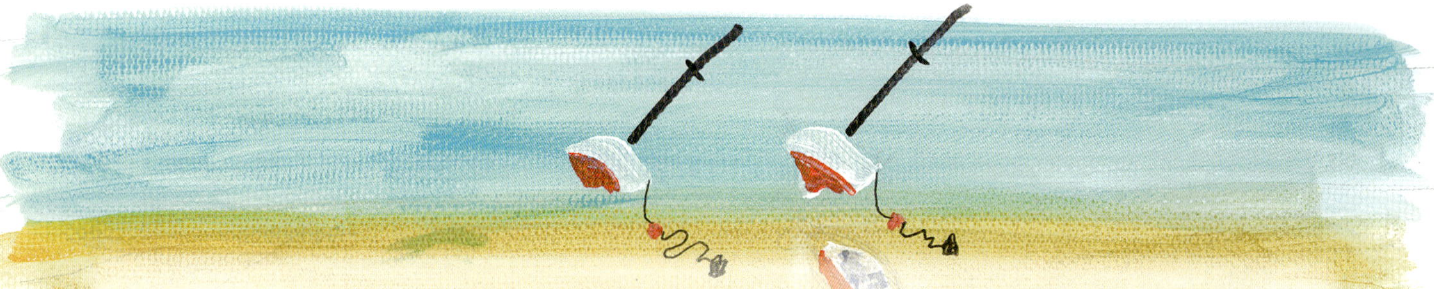

The tides are mainly caused by the moon's gravitational pull tugging on the oceans of the earth. The tug of the moon makes a big bulge of water on the side of the earth facing the moon, with another bulge of water on the side of the earth away from the moon. And because the earth spins around once every 24 hours, New England passes through a bulge of water twice per day. As we pass though a bulge of water, the water rises, then falls—that is the tides. Because we know the timing of the earth's and moon's orbiting and spinning around, we can make schedules of the tides that can tell us months in advance when high or low tide will come and go.

Down Time's quaint stream
Without an oar,
We are enforced to sail,
Our Port—a secret—
Our Perchance—a gale.
What Skipper would
Incur the risk,
What Buckaneer would ride,
Without a surety from the wind
Or schedule of the tide?

—Emily Dickinson
(New England poet, 1830–1886)

All along the coast, fog bells warn sailors to stay away from the rocks—but there are some very interesting rocks and fossils along the coast and under the waves!

You might find some beach glass from broken antique bottles, polished by the waves and sand.

We might think of New England as ending where the land meets the Atlantic Ocean, but there is more of New England under the sea.

In fact, when sea level was much lower thousands of years ago, because lots of ocean water was locked up as ice in the huge glaciers, animals roamed out in places that are now covered with water—a hundred or more miles offshore from where the coast is now.

"O father! I hear the church-bells ring,
Oh say, what might it be?"
"Tis a fog-bell on a rock-bound coast"—
And he steered for the open sea.
—Henry Wadsworth Longfellow

Offshore, under the sea, the rocks of New England continue east until the rocks of the continent end. There they slope down much deeper to where the rocks of the oceanic crust begin. The steep drop is called the continental slope. Much of the submerged land between the present New England coast and the continental slope is the vast Gulf of Maine. The Gulf of Maine has lots of underwater surprises. There are lobsters and crabs, fish and starfish, snails, and many other kinds of life under the surface of the sea.

Under the Atlantic there are many fascinating rocks and fossils, too—hidden beneath the sea. There are ancient river valleys, carved by rivers on once dry land, that were covered up by water when the sea level rose as glacial meltwater returned to the ocean. There are fossil bones and teeth of Mastodons—extinct, big, hairy, elephant-like animals that roamed parts of the Gulf of Maine when the sea level was lower during the Ice Age (tens of thousands of years ago). Sometimes fishing nets dragging the seafloor will bring up a mastodon's tooth!

Offshore of New England, deep under the Atlantic Ocean, there is a chain of volcanic "sea mounts" that were once active volcanoes just like the Hawaiian Islands. Like the Hawaiian Islands, they formed in a line because the plate that New England is riding on slowly moved over a "hot spot" within the earth (an area of hot magma that rises up). As New England moved over the stationary hot spot, the heat melted some of the moving plate's rock, which became granite or erupted from volcanoes. As the plate moved to the west, the hot spot created a string of volcanoes under the sea. You can see how this happens if you hold a thick magic marker under a paper towel—hold the marker still under the paper, and then move the paper over the marker and you can see the trail of ink it leaves.

So from the Berkshire Hills of western New England, all the way to where the continent ends beneath the Atlantic Ocean, we have seen what's under New England. There are ancient continents and volcanic island chains that moved across former seas to collide together and form New England.

So one day, if you are on the New England coast and get up early to watch the sunrise over the ocean, maybe you'll see the bright orange sun behind a bank of clouds on the horizon, making it look like some fiery volcanic island chain is approaching.

Don't worry, it will not be a volcanic island chain, or a micro-continent, that is going to collide with New England . . . but you will be standing on one that already has.

Nay, be a Columbus to whole new continents and worlds within you, opening new channels, not of trade, but of thought.

—Henry David Thoreau

Now you know what's under New England.
Know what's under its old and familiar landscape.
Know its granites that were once slow-cooling magma deep below ground,
And the volcanic rocks from fast-cooling lava that pushed up to the surface,
You know New England's sedimentary rocks from the floors of ancient oceans,
And its metamorphic rocks baked when New England was crunched between continents,
And you know New England's surface was sculpted by icy cold glaciers.

You know more about what's under New England, on- or offshore,
And the story of New England's rocks and fossils from bottom to top.

But there is always more to learn of this story. And much more to discover yourself. So after the sun rises over old New England, go out and explore the land. Look closely and carefully at the rocks.

You will find some pieces of once far-off lands,
You could find rocks from a fiery volcano,
You might hold a fossil from some ancient sea in your hand.
Who knows? You just might find some gold in the rocks of New England.

But best of all,

. . . you'll have fun just looking.

New England Geological Exploration Field Notes

Explorer: ______________________________

Date: __________________________________

Location: _____________________________

Observations:

List of Poems and Quotation Sources

Page 2: Henry Wadsworth Longfellow, from "A Gleam of Sunshine"
Page 8: Henry David Thoreau, from *Walden*
Page 19: Edna St. Vincent Millay, from "Renascence"
Page 25: Ralph Waldo Emerson, from "Musketaquid"
Page 28: Ralph Waldo Emerson, from "Song of Nature"
Page 33: Henry Wadsworth Longfellow, from "The Building of the Ship"
Page 35: Frank Dempster Sherman, from "Nantucket"
Page 38: Robert Frost, from "Our Singing Strength"
Page 42: Amy Lowell, from "Lilacs"
Page 45: Henry David Thoreau, from *Walden*
Page 47: Robert Frost, from "Once by the Pacific"
Page 49: Emily Dickinson, from "Down Time's Quaint Stream"
Page 51: Henry Wadsworth Longfellow, from "The Wreck of the Hesperus"
Page 55: Henry David Thoreau, from *Walden*

CHARLES FERGUSON BARKER, a geologist with a graduate degree from Boston University, has done extensive fieldwork across New England. He now teaches geology at Wayne State University and works for an environmental consulting firm in Detroit. He is the author and illustrator of *Under Michigan* (2005) and *Under Ohio* (2007).